Lifescience – Concepts analysed in a different way

VOLUME 2

DARANI VASUDEVAN

Preface to the title....

The contents of the book are prepared in such a way that it will make the readers to analyse the topics we usually read in any standard or preferred textbooks in a different manner like comparing the theoretical contents with life events or any research observations. The idea behind this is the questions asked in part C of CSIR UGC NET examination conducted widely throughout India for the post of Assistant Professor and Junior Research Fellow. This book is filled with concepts regarding various subdivisions of Life Science like systematic, ecology, biochemistry, cell biology, genetics, molecular biology, immunology, physiology and evolution. Most of us did not have the practice of analysing or comparing the idea behind the results we obtain in our research and projects with the textbook contents. The main idea of Part C is to make us relate both theory and practical. This book will help the readers to prepare for Part C in a better way than to simply read text books that is not a fruitful idea to succeed in that section of CSIR UGC NET examination. This is the second volume o the series. Hope you all enjoy and find it useful!!!!

V.Darani M.Sc., M.Phil., (SET)

# Concepts related to Cell Biology, Genetics and Molecular Biology

Some Terms and concepts Explained

INTRACELLULAR SORTING

Intracellular sorting requires specific mechanism for transport of proteins in and out of each compartment.

PULSE CHASE TECHNIQUE

In this technique, the cells are exposed to a labelled compound (pulse) and the cellular processes occurring over time are examined. The cellular processes of the the same compound in an unlabelled form (chase) is also noted. The commonly used label is the radioactive element

ROBERTSONIAN TRANSLOCATION

Robertsonian Translocation (ROB) is a rare form of chromosomal rearrangement in which the chromosomes break at their centrosomes and the long arms fuse to form a single large chromosome with a single centromere. In humans ROB is noted in 5 acrocentric chromosome pairs viz., 13,14,15,21 and 22. ROB is also referred to as whole arm translocation or centric fusion translocation. Dividing of chromosome into large arm containing majority o genetic content and a short arm with a much smaller proportion of genetic content takes place. Short arms too join and form a small reciprocal product which typically contains only non essential genes. These genes also occur elsewhere in the genome but are usually lost within a few cell divisions. In humans, ROB joins the long arm of chromosome 21 with the long arm of chromosome 14 or 15 resulting in unbalanced trisomy 21 (cause down syndrome) and trisomy 13 (patau syndrome).

AMES TEST

Ames test uses bacteria to determine whether a given chemical can cause mutation in the DNA of the test organism or not. The test employs several strains of bacterium *Salmonella typhimurium* that carry mutations in genes involved in histidine synthesis. These strains are auxotrophic mutants i.e. they need histidine for their growth but cannot produce it. The mutation causing capability of the substance is tested which results in the return to "phototrophic" state, so that the cells can grow in a histidine free medium. The tester strains are specially constructed to detect the frameshift or point mutations in the genes responsible for the synthesis of histidine. Larger organisms have metabolic processes which are capable of turning a non mutagenic chemical mutagenic one. In some cases, rat liver extract is usually added to stimulate the process.

SOME RNA TYPES AND THEIR FUNCTIONS:

- Sn RNAs function in a variety of processes like splicing of pre-mRNA
- Si RNAs direct degradation of selective mRNAs and turn of gene expression.
- mi RNAs block translation of selective mRNAs andregulate gene expression
- Sno RNAs are used for processing and chemical modification o tRNAs.

NOD GENES

Nod genes are lipo chitooligosaccharides. Nod genes are broadly classified into three types namely regulatory, common and host specific. Nod A, B, C are common nod genes which are present in all *Rhizobium sp*. The function o Nod A is N-acylation of aminosugar backbone. Nod B functions as the chitooligosaccharide deacetylase and Nod C acts as N-acetylglucosaminyl transferase. Nod D is regulatory gene that encodes DNA binding proteins which are responsible for the activating transcription in other nod operons. Nod P and Q are host specific Nod genes that are needed for the production of nodulation (nod) factors in *Rhizobium meliloti*.

DNA TRANSPOSONS

DNA transposons are able to move in the DNA of an organism through a single or double stranded DNA intermediate. The transposase enzyme is required by the system. This enzyme catalyzes the movement of DNA from its current location in the genome to a new location. In the process of transposition, the enzyme "cuts" the DNA segment and "Paste" it at some other point.

RETROTRANSPOSONS

Retrotransposons are transposons via RNA intermediate. These are genetic elements capable of amplifying themselves in a genome. The retrotransposon exhibit a replicative mode of through RNA intermediate and rapidly increases the copy number of elements thereby increasing the genome size. It can induce mutation by inserting near or within the genes.

ABC MODEL OF FLORAL DEVELOPMENT IN ARABIDOPSIS

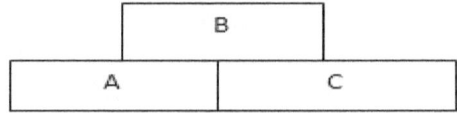

The class A genes are Apetala 1 and Apetala 2. Class B genes are Apetala 3 and Pistillata (P_1). Class C gene is Agamous (Ag). Sepal formation is controlled by Class A genes. Class A and B together are responsible for petal formation and Class B and C control the formation of stamens and Class C alone is responsible for carpels formation. Class A and Class C are reciprocally antagonistic. The expression of mutants are as follows:

	Class A mutant	Class B mutant	Class C mutant
1st whorl	Carpel	Sepal	Sepal
2nd whorl	Stamen	Sepal	Petal
3rd whorl	Stamen	Carpel	Petal
4th whorl	Carpel	Carpel	Sepal

The control of flowering is a complex process involving several key regulatory genes. Following are statements related to flower development

- Two major types of genes regulate floral development: meristem identity genes and floral organ identity genes.
- The important genes in Arabidopsis that play key regulatory role in meristem identity are: APETALA 1, LEAFY and SUPPRESSOR of CONSTANS 1.
- The genes that determine floral organ identity were discovered as floral homeotic mutants

HOMEOTIC GENES

The genes that regulate the development of anatomical structures are named as Homeotic genes. Mutation in these genes results in displacement of body parts. Hox and Para Hox genes are associated with segmentation.

COP1 PROTEINS

COP1 are proteins that inhibit photomorphogenesis. During dark hours COP1 is present in the nucleus but in the presence of light its occurrence is concerned in the cytoplasm. COP1 contains a ring finger like structure which is a feature of many E3 ubiquitin ligases, that are responsible for targeting proteins in 26s proteosome mediated degradation. It interacts with HY5 and in the dark HY5 protein levels show a decline that is dependent on COP1 and CNS. Hence it could be understood that in the dark, Cop1 targets HY5 for degradation by CNS.

SOME CELL TO CELL ADHESION PROTEINS AND THEIR FUNCTIONS

- Selectin are lecitins that mediate a variety of transcient, cell-cell adhesion interactions in the blood stream

- Ig-Sugar family are proteins that contain extracellular Ig like domains and are mainly involved in the fine tuning of cell-cell adhesive interactions during development and regeneration.
- Cadherin mediates Ca^{2+} dependent strong homophilic cell-cell adhesion.
- Integrin are transmembrane cell adhesion proteins that acts as extracellular matrix receptors.

Reasoning of questions related to cell biology, genetics and molecular biology asked in CSIR UGC NET

The principal pathway for transport of lysosome hydrolases from the trans golgi network (pH 6.6) to the late endosomes (pH 6.0) and the recycling of M6P (Mannose 6 phosphate) receptors back to golgi depends on the pH difference between those two compartments. From what you know about M6P receptor binding and recycling and the pathways of delivery of material to lysosomes, if the pH in late endosomes was raised to 6.6, the receptor would not release hydrolase and could not be recycled back to trans golgi network.

Reason

M6P receptors are transmembrane glycoproteins responsible for targeting enzymes to lysosomes. The targeted enzymes are lysosomal enzymes that are synthesized in the Rough Endoplasmic reticulum. M6P receptors bind with the newly synthesized lysosomal hydrolases in the trans Golgi network (TGN) and deliver them to pre lysosomal compartments. The binding of ligands is most effectively at pH 6-7. The receptor binds to the lysosomal enzymes in trans golgi and release them in the acidified environment of endosome. Once the enzyme has dissociated, the phosphate taq is removed from the enzyme. The dissociation occurs at the pH 6.0. After dissociating the M6Ps are pached into vesicle that bud off the late endosome and return to the trans Golgi complex.

Life science II

Cells that grow and divide in a medium containing radioactive thymidine covalently incorporate the thymidine into their DNA during S phase. Consider a simple experiment in which cells are labelled by a brief (30 minutes) exposure to radioactive thymidine. The medium is then replaced with one containing unlabeled thymidine and the cells grow and divide for some additional time. At different time points, after replacement of the medium, cells are examined under a microscope. Cells in mitosis are easy to recognize by their condensed chromosomes and the fraction of mitotic cells that have radioactive DNA can be estimated by autoradiography and plotted as a function of time after the thymidine labelling. There was a fall in curve at 20 minutes then there was a rise in curve after 20 minutes. This corresponds to cells in apoptotic phase.

Explanation

Mitosis and cytokinesis together constitute mitotic M phase. During mitosis, the chromosome that has already been duplicated, condense and attaches to spindle fibres which pull one copy of each chromosome to opposite sides of the cell. As a result two genetically identical daughter nuclei are formed. The rest of the cell then divides completing the formation of two daughter cells. Errors may occur which include production of three or more daughter cells resulting in Tripolar mitosis or multipolar mitosis (i.e. direct cell triplication or multiplication). Other errors include apoptosis or mutation leading to cancer cells. The fall in curve at 20 minutes indicates the cells in apoptosis due to changes in medium. After getting steady it started to rise.

Inorder to study the role of telomeres in DNA replication, genetically engineered mice were prepared, where the gene for telomerase RNA was knocked out. When cells from these knockout mice were taken and cultured in vitro, they proliferated even after 100 cell divisions which is quite unlikely in the case of human cells because mice have very long stretch of telomere sequence compared to that of human.

Explanation

The ends of the chromosome are named as telomeres. They have tandem repeats of G rich sequence (TTAGGG). With the aid of DNA polymerases and primers, the enzymes replicate in the terminal base of DNA molecule. The chromosomes would get shorten with the removal of terminal RNA primers at each round of replication (end replication problem). This loss is compensated by the telomeres. Capping is a mechanism for replicating and maintaining chromosome termini. Telomeres of laboratory mice are 5 to 20 times longer than in human but their life span is 30 times shorter.

A non enzymatic viral protein X was found to be inducing a cellular gene promoter activity. Although no invitro DNA binding activity could be identified with X protein, it was found to be co-recruited on the cellular promoter along with a cellular transcription factor invitro. The best interpretation from the above is that X physically interacts with the transcription factor.

Explanation

It is given that X induces gene promoter activity and shows no binding activity and is also found to be co-recruited with cellular transcription factor hence, X physically interacts with the transcription factor.

During elongation step of protein synthesis, translocation moves the mRNA and the peptidyl t-RNA by one codon through the ribosome. Translocation in E.coli involves GTP and EF-G. However invitro translocation can take place independent of GTP and EF-G. From this it could be concluded that the translocation activity is inherent in ribosomes, however the rate of translocation invivo is enhanced significantly in presence of GTP and EF-G.

Explanation

During protein synthesis, elongation is followed by a translocation step in which mRNA and tRNA are advanced by one codon. This step is catalyzed by elongation factor (G)-EFG, a guanosine triphosphatase (GTPase) along with a rotation between the two ribosomal subunits. Thus, translocation is catalyzed by the two which are not needed in vitro because the artificially added conditions enhance the process.

In cells having G-protein coupled receptor, inhibition of protein kinase A by siRNA technology led to diminished transcription of androgen binding protein (ABP) and CREB protein. Addition of cAMP which is a secondary messenger will lead to no change in transcription level.

Explanation

Androgen Binding Protein (ABP) in sertoli cells causes the accumulation of androgen. When ABP binds to its receptor accumulation of cAMP takes place. cAMP response element binding protein (CREB) binds to certain DNA sequences called CRE and increases or decreases the transcription of the downstream genes.

Binding of a ligand to a cell surface receptor activates an intracellular signal transduction pathway through the sequential activation of four protein kinases. In the human cell line A, these kinases are held in a signalling complex by a scaffolding protein whereas in another cell line B, these kinases are freely diffusible. The possibilities are speed of signal transduction will be higher in cell A. Possibility of cross linking with other signal transduction pathways will be lesser in cell A. The potency of spreading signal through other signalling pathways will be higher in cell B.

Explanation

It is given that binding of a ligand to cell surface receptor activates intracellular signal transduction mediated by four protein kinase. In the cell line A four protein kinases are held by scaffolding protein. The signal transduction is high and hence the cross linking with other pathways is not possible. In the case of cell line B, the four protein kinases are left free. Hence the signal transduction is low and cross linking is possible.

Typical morphological defects are routinely used in genetic screens to identify novel genes in signal transduction pathways. Triple response morphology of seedling has been used to decipher the ethylene signalling pathways.

Explanation

Mutants displaying a constitutive response (eto-1) produce 40 times more ethylene than the wild type. When ethylene biosynthesis or ethylene action was inhibited, the morphological defects in etiolated eto1-1 seedling reverted to wild type. In the absence of ethylene mutants failed to display the apical hook and also showed reduced ethylene production. Even in the presence of exogenous ethylene, the elongation of hypocotyls and root of etiolated seedlings were inhibited but the apical hook was still not observed. Mutants that were insensitive to ethylene produced increased levels of ethylene and displayed hormone insensitivity in both hypocotyl and root responses. They exhibited apical hook. Each of the "triple responses" mutants has an effect on the shape of the seedling and also on the production of the hormone. These mutants thus proved to be useful tools for analysing the mode of ethylene action in plants.

In an experiment on transposition in eukaryotic system, an intron was cloned within a transposable element and allowed to transpose from a plasmid to genome DNA. The intron was found to be absent in the transposable element in its new location. It is retroposon.

Explanation

Retroposons are repetitive DNA fragments which are inserted into chromosomes after they are reverse transcribed from RNA. Transposons are DNA sequences that can change its position within a genome. It is of two types, retrotansposons and DNA transposons. Transposition is the mechanism and it occurs in two ways: Autonomous and non- autonomous. In autonomous the transposons move by themselves. In the case of non autonomous mode, they need another transposing element (TE) to move. They lack transposase or reverse transcriptase which aids them in transposition. Example: Retroposon.

There are two mutant plants. One shows taller phenotype than wild type, whereas the other has the same height as the wild type. When these two mutants were brought in together by genetic crosses, the double mutant displayed even taller phenotype than the tall mutant plants. This genetic interaction is called synergistic interaction

Explanation

Antagonistic interaction involves defensive strategies that make use of certain chemical and physical deterrents. It is the best interaction for obtaining antidote for poison. The sum of the effects of chemicals is involved in reaction is termed as additive effects. Synergistic interaction occurs when sum of the effects of chemicals is more than effect of each chemical individually. This creates a dangerous situation because each chemical is designed to work well on its own. This interaction is like pairing two really strong superheroes. Individually they are both strong and if combined, their powers are overwhelming.

Plasmids are self replicating small circular DNA elements in a bacterial cell that can be said to have a stable symbiotic existence with the host cell. They often carry genes useful for the host. Copy up mutation that increases the rate of plasmid replication per host cell cycle is a potential threat to the evolution and stability of the symbiotic coexistence.

Reason

The number of host cells increases as a result of rapid replication leading to tumour.

In an experiment that has continued for more than 50 years, corn has been propagated by breeding only from plants with the highest amount of oil in the seeds. The average oil content is now much greater than any of the plants in the original population. The hypothesis that could be postulated from the above observation are

- ➢ Mutation has occurred resulting in increased the oil content in seeds.
- ➢ The breeding led to increased frequency of alleles at multiple loci, so that new combinations of genes for even higher oil content were formed.

While attempting to create a disease model of poliomyelitis in mice, it was found that mice cannot be infected with the said virus. Since human beings are susceptible to this viral infection, the kind of transgenic mice that should be generated to have a transgenic mouse model that can be infected with polio is a mouse expressing human receptor gene which makes cell surface protein for docking and internalization of polio virus.

Reason

A gene encoding a protein acting as a cellular receptor for poliovirus is identified from human cells. This human gene when made to express in mouse cells made them susceptible to poliovirus infection. The gene was identified to encode a trans membrane glycoprotein called poliovirus receptor (PVR) later renamed CD155.

The correct sequence of events in a next generation sequencing technology based whole genome sequencing project is

- ♠ DNA extraction
- ♠ Shearing
- ♠ Adapter ligation
- ♠ Library amplification
- ♠ Sequencing
- ♠ Assembly
- ♠ Finishing
- ♠ Annotation
- ♠ Submission to Gene bank.

An investigator discovers a new receptor for a known ligand and wanted to identify the binding partner of the receptor (i.e.) its co receptor. The anti receptor antibody is not available but anti GFP antibody is available. The strategy is most likely to identify the co-receptor is the receptor is cloned as a fusion protein of GFP and expressed in stimulated cells. The immune precipitated complex obtained by anti GFP antibody was analysed by LC-MS/MS.

Reason

Co receptor is a cell surface receptor which binds to signalling molecule in addition to a primary receptor. It facilitates ligand recognition and initiates biological processes like entry of pathogen into host cell. Since Co receptors are proteins, they maintain a 3D structure. The identification of co receptor is carried out in the same way as other proteins viz., using GFP (Green fluorescent protein).

Using FRAP (Fluorescent Recovery after Photo bleaching) techniques, diffusion coefficient of three integral membrane proteins M_1, M_2 and M_3 in a kidney cell is calculated as 1µm/s, 0.05 µm/s and 0.005 µm/s respectively. Considering fluid mosaic nature of biological membrane and relationship of structural organization of integral membrane protein with diffusion coefficient, the protein that has highest number of integral membrane domain is M_3.

Explanation

As per the given data it could be noted that the three different proteins have different diffusion coefficients from the same cell i.e. the membrane fluidity remaining the same in all three cases. The mobility of protein is affected by structural organization of integral proteins. The integral membrane domains of proteins may either associate with cytoskeletal proteins or act as receptor and interact with cytoplasmic proteins because of which such proteins will have reduced mobility. This implies that proteins with higher number of integral domain will have reduced diffusion capability in a membrane.

When the cells enter mitosis, their existing array of cytoplasmic microtubules has to be rapidly broken down and replaced with the mitotic spindle, which pulls the chromosome into daughter cells. The enzyme katanin is activated during the onset of mitosis and chops microtubules into short pieces. The possible fate of microtubule created by katanin will be depolymerisation.

Reason

Katanin is a microtubule responsible for dividing AAA protein (ATPase associated with diverse cellular activities). This dividing process is regulated by nucleotide exchange factors called protective microtubule associated proteins (MAPs) that can exchange ADP with ATP. Katanin mediated microtubule dividing is an important step in mitosis and meiosis. It divides microtubules at the mitotic spindles when disassembly is required to segregate sister chromatids during anaphase.

In recent years, genome wide transcription study using high throughput sequence analysis has revealed some novel results that include

- *Presence of RNA polymerase in both intra and intergenic regions of the genome*
- *Existence of non- coding RNAs generated from mRNA coding genes.*
- *Existence of sense and antisense transcripts generated from the promoter and untranslated region of many annotated genes.*

The possible interpretations of the above results are:

- *Binding of RNA polymerase to non promoter regions of the genome leads to the generation of various non coding regulatory RNAs.*
- *Through bidirectional transcription of sense and antisense transcripts are generated from the promoter and untranslated regions of protein coding genes.*

Explanation

The double stranded DNA is opened up by RNA polymerases such that one strand acts as template for RNA synthesis. When two RNA polymerases initiate on the same point of DNA but on opposite strands, transcription of each strand will occur but RNA polymerase move away from each other as a result divergent transcription and hence the transcripts would not overlap. Consider two polymerases one initiating on a gene and the other initiating within a gene resulting in a convergent transcription. Here the polymerases would be moving towards each other but transcribing opposite strands. Since the strands would be complementary where the polymerase passes each other, this would give rise to antisense transcripts. These two processes are simply referred as bidirectional transcription.

Vascular endothelial (VE)- Cadherin is an important cell adhesion molecule for endothelial cells. Endothelial cells that are unable to express VE- Cadherin still can adhere to one another via N-Cadherin (neural cadherin), but these cells do not survive. The appropriate reason for this is VE- cadherin acts as co-receptor for VEG F (Vascular endothelial growth factor)- mediated signal transduction in endothelial cells.

Reason

VE cadherin is a cell adhesion molecule for endothelial cells. Some endothelial cells are unable to express VE cadherin but adhere to each other by neutral cadherin but do not surviveThe function of VE cadherin is to maintain vascular integrity. VE cadherin is responsible for organization of stable vascular system during the development of embryo. In adult it controls vascular permeability and inhibits unrestrained vascular growth. It acts through mechanisms like reshaping and organization of endothelial cell, cytoskeleton and modulation of gene transcription. VE cadherin is calcium dependent.

An important role of Fas and Fas ligand is to mediate elimination of tumor cells by killer lymphocytes. In a study of 35 primary lung and colon tumors, half of the tumors were found to have amplified and overexpressed a gene for a "secreted protein" that binds to Fas ligand. This is the main reason for survival of these tumours. Cells by this "secreted Fas ligand binding protein" may be attributed to its Decoy receptor activity.

Reason

A decoy receptor is a receptor responsible for recognizing and binding of specific growth factors or cytokines efficiently, but is structurally unable to signal or activate the intended complex. It acts as an inhibitor, binding a ligand and keeping it away from binding to its receptor. Decoy receptors participate in common methods of signal inhibition and are also abundant in malignant tissue.

Maturation promoting factor (MPF) controls the initiation of mitosis in eukaryotic cells. MPF kinase activity requires cyclin B. Cyclin B is required for chromosome condensation and breakdown of nuclear envelope into vesicles. Cyclin B degradation is followed by chromosome decondensation, nuclear envelope reformation and exit from mitosis. This requires ubiquitination of a cyclin destruction box motif in cyclin B. RNase-treated Xenopus egg extracts and sperm chromatin were mixed. MPF activity increased with chromosome condensation and nuclear envelope breakdown. However this was not followed by chromosome decondensation and nuclear envelope reformation.

Reason

The lack of chromosome decondensation is because cyclin B lacking the cyclin destruction box had been over expressed.

In order to study the transcription factor TFIIH, it was cloned from a large number of human subjects. Surprisingly, the subjects having mutation in TFIIH, also showed defects in their DNA repair system.

Reason

This is because in mammalian system, TFIIH plays an active role in transcription coupled DNA repair process.

If wingless RNAi is expressed in wingless expressing cells from the stage where this gene initiates its expression in a developing Drosophila embryo, the posterior compartment of each future segment will get affected.

Explanation

Wnt/wingless (wg) signalling pathway plays a variety of role in animal development. Regulation of Armadillo (arm) protein levels through ubiquitin mediated degradation plays the main role in wingless signalling. In both plants and animals RNA interference (RNAi) is mediated by small RNAs which are of 21-23 nucleotides in length is used for regulation of target gene expression at multiple levels through partial sequence complementarities. The two target genes namely vestigial and shot gun of wnt/wingless signal transduction pathways aid in controlling the columnar shape of Drosophila wing disc cells. Wing disc cell is a sac-like structure comprising of columnar epithelium or disc proper cells (DP), the cuboidal marginal cells (MC) and the overlying squamous cells (SC). Regulatory lines are employed in the integration of patterning information at several stages during *Drosophila* wing development. The lines prevent wing accumulation in wing primodium, confining its expression to the peripodial epithelium. The wing alteration is not caused by the expression of Bowl- RNAi in DP cells of wing pouch whereas the expression of Lin-RNAi caused a huge reduction in wing size, indicating that the development was severely compromised.

Formation of digits and sculpting the tetrapod limb requires death of specific cells in the limb in a programmed manner.

Explanation

The canonical wnt pathway is the widely studied Wnt signalling pathway that controls the expression of gene by causing β-accumulation when wnts are present. In the case of absence of wnts the nuclear β- catenin levels are reduced. The Non canonical wnt signalling pathway on the other hand is β- catenin independent. During the limb initiation stage the growth and formation of limb buds are induced by the lateral plate mesoderm at the fore and hind limb. The underlying limb mesoderm induces the Apical ectoderm ridge (AER) which forms the distal tip of the ectodermal pocket that runs along the antero-posterior axis of the limb bud. This is crucial for proximo distal limb outgrowth. As the limb development initiates, the limb buds forms as a result of interplay between fibroblast growth factor (FGF) and wnt signalling, both acting downstream of the T-box transcription factors Tb×5 and Tb×4 in the developing forelimb and hindlimb respectively. BMP is concerned with skeletal formation. The antagonistic inhibitor of wnt siganalling pathway is the Dickkopf/Wnt signalling pathway inhibitor 1 (DKK-1).

Somatic recombination was caused by mild exposure to radiation on flies heterozygous for a given allele during specific stages of development and the individuals were allowed to develop. Such individuals are likely to have:

- Clones of homozygous cells in heterozygous body
- Twin spots (i.e.) patches of mutant cells and homozygous wild type cells in heterozygous body.

Explanation

The powerful tool used in the study fruit flies are the genetic mosaics, where they are created through mitotic recombination. Formerly mosaics were created by irradiating flies that are heterozygous for a particular allele with X-rays. This induces break in the double strand DNA. When repaired this could result in a cell homozygous for one of the two alleles. This cell would result in a patch or "clone" of cells mutant for the allele being studied after further rounds of replication. Twin spotting in D-melanogaster lead to the discovery of mitotic recombination.

Cellular level of tumour suppressor protein P53 is maintained by the ubiquitin ligase protein, Mdm2. Over expression of Mdm2 was found to convert a normal cell into cancer cells by destabilizing P53. Another protein 19^{ARF} inhibits the activity of Mdm2 thus stabilizing P53. Loss of $P19^{ARF}$ function also converts normal cells. Based on the above information what could be concluded?

Answer

It is very clear from the given data that Mdm2 is an oncogene but 19^{ARF} is a tumor suppressor gene.

As topoisomerases play an important role during replication, a large number of anticancer drugs have been developed that inhibit the activity of these enzymes. Statements true about topoisomerases as a potential anticancer drug target:

- As cancer cells are rapidly growing cells, they usually contain higher levels of topoisomerases.

19 | Life science II

- ♠ The transient DNA breaks created by topoisomerases that are usually converted to permanent breaks in the genome in the presence of topoisomerase targeted cells.
- ♠ As cancer cells often have impaired DNA repair pathways, they are more susceptible towards topoisomerases targeted drugs.

The Reason for above conclusion

Topoisomerases play role in overwinding or unwinding of DNA.

Each Aminoacyl t-RNA synthase is previously able to match an aminoacid with the tRNA containing the correct corresponding anticodon. Most organisms have 20 different tRNA synthetases, however some bacteria lack the synthetase for charging the tRNA for glutamine (tRNA^{Gln}) with its cognate aminoacid. How do these bacteria manage to incorporate glutamine in their proteins?

Answer

In these bacteria, the aminoacyl tRNA synthetase specific for tRNAglu also charges tRNA^{gln} with glutamate. A second enzyme then converts the glutamate of the charged tRNA^{gln} to glutamine.

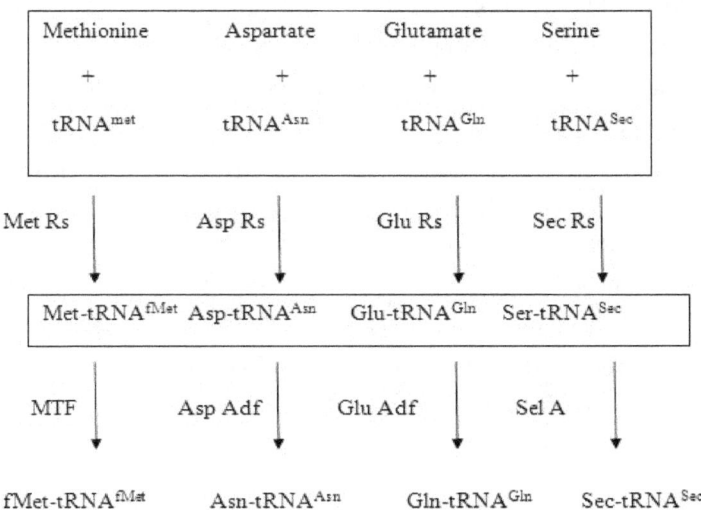

fMet-tRNA^{fMet} is involved in translation initiation. Asn-tRNA^{Asn} and Gln-tRNA^{Gln} are involved in translation elongation nad aminoacid metabolism. Sec-tRNA^{Sec} is involved in missense suppression.

A eukaryotic cell undergoing mRNA synthesis and processing was incubated with 32P labelled ATP, with the label at the β position. 32P will not appear in the mature mRNA under any circumstances because β and γ phosphates are released during transcription

Reason

β and γ phosphate groups are removed as pyrophosphate.

Four different mutant lines showing the similar phenotype were identified from a genetic screen. When genetic crosses among these mutants were carried out, the first mutant was found to complement the second, third and fourth mutant lines. Ahowever no other complementation was observed. The numbers of complementation groups the four mutant lined belong are 2.

Reason

Using the cis-trans or complementation test, the functional allelism of any two recessive mutations can be determined experimentally. Trans heterozygotes are the organisms or cells having two mutations in the trans or repulsion configuration. If the wild type phenotype is present in trans heterozygote then the two mutations are in different units of function i.e.two different genes. In such case, the two mutations are said to complement each other.

During transgenesis, the location of the genes and their number integrated into the genome of the transgenic animal are random. It is often necessary to determine the copy number of genes and their tissue specific transcription. The possible methods used for identification are?

Answer

The most commonly employed methods to determine the copy number of genes and their tissue specific transcription are southern blot hybridization and reverse transcriptase PCR.

In order to clone a eukaryotic gene in pBR322 plasmid vector, the desired DNA fragment was produced by Pst I cleavage and incubated with Pst I digested pRR322 (Pst I cleavage site lies within the ampicillin resistant gene) and ligand. Mixture of ligated cells was used to transform E.coli and plasmid containing bacteria were selected by their growth in tetracycline containing medium. Which type of plasmids will be formed?

Answer

A Circular pBR322 plasmid having the target gene and found to be resistant to tetracycline, a recircularised pBR322 resistant to both ampicillin and tetracycline and a concatemarised pBR322 resistant to both ampicillin and tetracycline.

Inorder to ensure that only fully processed mature mRNAs are allowed to be exported to cytosol, pre-mRNAs associated with SnRNPs this, an experiment was performed where a gene coding a pre-mRNA with a single intron was mutated either at the 5' or 3' splice sites or both the splice sites. It was observed that, pre-mRNA having mutation at both the splice sites will be exported to cytosol. Pre-mRNA mutated at either 3' or 5' splice sites will be exported to cytosol because of the absence of bound snRNPs

Explanation

As a result of mutation to pre mRNA gene snRNP is prevented from binding to it and hence it is not retained in the nucleus.

In mammals CG rich sequences are usually methylated at C, which is a way for marking genes for silencing. Although the promoters of housekeeping genes are often associated with CPG islands yet they are expressed in mammals. It is because unlike within the coding region of a gene, CG rich sequences present in the promoters of active genes are usually not methylated.

Explanation

The CPG islands typically remain unmethylated in promoters of housekeeping genes.

In an experiment, RBCs were subjected to lysis and any unbroken cells were removed by centrifugation at 600g. The supernatant was taken and centrifuged at 100,000 g. The pellet was extracted with 5M NaCl and again centrifuged at 100,000g. The protein that is present in the supernatant would be spectrin.

Explanation

The integral membrane proteins band 3 glycophorin and GPCR are difficult to remove by 5M NaCl salt treatment.

Inorder to study the intracellular trafficking of protein 'A' it was tagged with GFP (A-GFP). Fluorescence microscopy showed that A-GFP co localizes with LAMP-1. In the presence of bafilomycin A, an inhibitor of H^+-ATPase, A-GFP does not co localize with LAMP1. Instead, it co-localizes with LC3 puncta. This is because autophagy is required for trafficking of A-GFP to lysosomes.

Explanation

Lysosomes are the home to Lysosomal associated membrane protein (LAMP) whereas LC3 puncta is present in autophagosome. During autophagy, the autophagosome fuses with lysosome resulting in the degradation of its contents. The protein A-GFP should always co localizes with LAMP1.. Since Bifilomycin inhibits the formation of lysosome, the protein A-GFP which is unable to co-localize with LAMP1 co-localizes with LC3in autophagosome.

In trypanosomes, a 35 base leader sequence is joined with several different transcripts making functional mRNAs. The leader sequence is joined with the other RNAs by the process of trans- splicing

Explanation

All RNAs are presented by the combined action of trans splicing and polyadenylation in trypanosomes. In trans splicing the small RNA molecule siRNA donates a leader sequence to all mRNAs.

Concepts related to Biochemistry and Microbiology

Key Notes

LIST OF PRECURSOR AMINO ACID AND ALKALOID CLASSES

Precursor amino acid	Alkaloid classes and examples
Ornithine	Pyrrolidine, Trophane, Pyrrozidine, Polyamines
	Cocaine, Atropine, Putrescine, Spermine, Spermidine
Leucine	Pyrrole
Lysine	Pyperidine, quinolizidine, Indolizidine
Tyrosine	Isoquinoline, Tetrahydroisoquinoline, benzyl tetrahydroisoquinoline, catecholamines
	Morphine, curarines, Papavarine, Nor adrenaline, Adrenaline
Tryptophan	Indole, Carbolines, Quinoline, Pyrrolindole, Indolamines
	Vindoline, Catharactine, Quinine, Capthotecin, Melatonin, Seratonin
Histidine	Imidazole
Phenylalanine	Ephedrine, Capsaicin
Anthranilic acid	Quinazoline, Quinoline, Acridine
Purine	Theobromine, theophylline, Caffeine
Geranylgeranyldiphosphate	Terpenoidic
Chloresterol	Steroidal, Solanin
Acetate	Piperidine
Nicotinic acid	Pyridine
	Nicotine

LIST OF PLANT DRUG AND THE SOURCE

Plant drug	Source Plant
Acetyldigoxin	*Digitalis lanata*
Adoniside	*Adonis vernalis*
Aescin	*Aesculus hippocastanum*
Aesculetin	*Frazinus rhychophylla*
Agrimophol	*Agrimonia supatoria*

Ajmalicine	*Rauvolfia serpentine*
Allantonin	*Many plants*
Allyl isothiocyanate	*Brassica nigra*
Anabesine (Skeletal muscle relaxant)	*Anabasis phyla*
Andrographolids	*Andrographis paniculata*
Anisodamine	*Anisodus tanguticus*
Anisodine	*Anisodus tanguticus*
Arecoline	*Areca catechu*
Asiaticoside	*Centella asiatica*
Atropine	*Atropa belladonna*
Berberine	*Berberis vulgaris*
Bergenin	*Ardisia japonica*
Betulinic acid	*Betula alba*
Bromelain	*Ananas comosus*
Caffeine	*Camellia sinensis*
Camphor	*Cinnamomum camphora*
Camptothecin	*Camptotheca acuminate*
(+)- Catechin	*Potentilla fragarioides*
Chymopapain	*Carica papaya*
Cissampeline	*Cissampelos pareira*
Cocaine	*Erythroxylum coca*
Codeine	*Papavar somniferum*
Colchicine amide	*Colchicum autumnale*
Colchicine	*Colchicum autumnale*
Convallatoxin	*Convallaria majalis*
Curcumin	*Curcuma longa*
Cynarin	*Cynara scolymus*
Danthron	*Cassia species*
Demecolcine	*Colchicum autumnale*
Deserpidine	*Rauvolfia autumnale*
Deslanoside	*Digitis lanata*
L-Dopa	*Mucuna species*
Digitalin	*Digitalis purpurea*
Digitoxin	*Digitalis purpurea*
Digoxin	*Digitalis purpurea*
Emetine	*Cephalis ipecacuanha*
Ephedrine	*Ephedra sinica*
Etoposide	*Podophyllum peltatum*

Galanthamine	*Lycoris squamigera*
Gitalin	*Digitalis purpurea*
Glaucarubin	*Simarouba glauca*
Glaucine	*Glaucium flavum*
Glasionine	*Octea glazionii*
Gossypol	*Gossypium sp.*
Hemsleyadin	*Hemsleya amabilis*
Hesperidin	*Citrus sp*
Hydrastine	*Hydrastis Canadensis*
Hyoscyamine	*Hyocyamus niger*
Irinotecan	*Camptotheca acuminata*
Kaibic acid	*Digenea simplex*
Kawain	*Piper methysticum*
Kheltin	*Ammi visage*
Lanatosides A,B,C	*Digitalis lanata*
Lapachol	*Tabebuia species*
a-Labeline	*Lobelia inflata (Indian tobacco)*
Menthol	*Mentha species*
Methyl salicilate	*Gaultheria procumbens*
Monocrotaline	*Crotolaria sessiliflora*
Morphine	*Papaver somniferum*
Neo andrographolids	*Andrographis paniculata*
Nicotine	*Nicotiana tabacum*
Nordihydroguaiaretic acid	*Larrea divaricata*
Noscapine	*Papaver somniferum*
Ouabain	*Strophanthus gratus*
Pachycarpine	*Sophora pschycarpa*
Palmitine	*Coptis japonica*
Papain	*Carica papaya*
Papavarine	*Papaver somniferum*
Phyllodulcin	*Hydrangea macrophylla*
Physostigmine	*Physostigma venenosum*
Picrotoxin	*Anamirta cocculus*
Pilocarpine	*Pilocarpus jaborandi*
Podophyllotoxin	*Podophyllum peltatum*
Protoveratrines A,B	*Veratum album*
Pseudo ephedrine	*Ephedra sinica*
Nor-pseudoephedrine	*Ephedra sinica*
Quinidine	*Cinchona ledgeriana*

Quinine	*Cinchona ledgeriana*
Qulsqualic acid	*Quisqualis indica*
Rescinnamine	*Rauvolfia serpentine*
Reserpine	*Rauvolfia serpentine*
Rhomitoxin	*Rhododendron molle*
Rorifone	*Rorippa indica*
Rotenone	*Lonchocarpus nicou*
Rotundine	*Stephauia sinica*
Rutin	*Citrus sp.*
Salicin	*Salix alba*
Sanguinarine	*Sanguinaria Canadensis*
Santoxin	*Artemisia maritma*
Scillarin A	*Urginea maritime*
Scopolamine	*Datura sp.*
Sennosides A, B	*Cassia sp.*
Silymarin	*Silybum marianum*
Sparteine	*Cytisus scoparius*
Stevioside	*Stevia rebaudiana*
Strychnine	*Strychnos nux vomica*
Taxol	*Taxus brevifolia*
Teniposide	*Podophyllum peltatum*
Tetrahydrocannabinol	*Cannabis sativa*
Tetrahydropalmatine	*Corydalis ambigua*
Tetrandrine	*Stephania tetranda*
Theobromine	*Theobromo cocao*
Thymol	*Thymus vulgarius*
Topotecan	*Camptotheca acuminate*
Trichosanthin	*Trichosanthes kirilowii*
Tubocurarine	*Chondodendron tomentosum*
Valapotriates	*Valveriana officinalis*
Vasicine	*Vinca minor*
Vinblastine	*Catharanthus roseus*
Vincristine	*Catharanthus roseus*
Yohimbine	*Pausinystalia yohimbe*
Yuanhuacine	*Daphne genkwa*
Yuanhuadine	*Daphne genkwa*

[Reference: C.K.Kokate, A.P. Purohit, S.B. Gokhale., (2002) Pharmacognosy 13th edition. Nirali prakashan]

Reasoning of questions related to Biochemistry and Microbiology asked in CSIR UGC NET

Where ΔG_{van} is the free energy of the Vanderwaals interaction, A and B are constants, r is the distance between the two non bonded atoms 1 and 2 and q_1 and q_2 are partial charges on the dipoles 1 and 2. In this relation the parameter it describes is the Dipole Dipole repulsion.

Explanation

The surrounding electron cloud influence with each other when two uncharged atoms are brought very close together. The repulsion of atoms and molecules at small distance can be explained as Vanderwaal's radius. In simplest case Vanderwaal's potential is isotrophic and written as

$$E = A/R^{12} - B/R^6 + q_1q_2/R$$

Here R is the distance between atoms and molecules and is the q_1 and q_2 partial charges.

The aminoacid alanine has high propensity to occur in helical conformation. The circular dichorism spectrum of an equimolar mixture of two 20- residue peptides, one composed of only L-alanine and the other only D-alanine is recorded in the region of 185-250 nm. In that case bands with identical negative and positive ellipticity will be observed.

Reason

The ellipticity of L-alanine is negative whereas that of D-alanine is positive. Since both are in equal composition, bands with identical positive and negative ellipticity were obtained.

Phosphorylation of serines as well as methylation and acetylation of lysines in histone tails affect the stability of chromatin structure above the nucleosome level and have important consequence for gene expression. The resulting changes in charge are expected to affect the ability of the tails to interact with DNA because DNA is a negatively charged one.

Explanation

An uncharged aminoacid can be converted to a negatively charged one through the phosphorylation of serine. The charge is not altered by the methylation of lysine whereas acetylation of lysine removes the positive charge thus leaving the modified lysine neutral. Both the introduction of a negative charged by phosphorylation of serine and removal of a positive charge by acetylation of lysine are expected to decrease the interaction of the negatively charged polymer histone tails of DNA.

According to the current model of alternative oxidase regulation, the factors that cause induction of alternative oxidase are presence of α keto acids (like pyruvate and glyoxylate) and cold stress.

Explanation

Two terminal oxidases namely cytochrome oxidase and cyanide insensitive alternative oxidase are present in mitochondria. Partitioning of electrons between the two pathways is regulated by the redox poise of theubiquinone pool and activation state of alternative oxidase. Ubiquinol is the substrate for alternative oxidase. The degree of reduction of quinone pool governs the activity of ubiquinol. The presence of pyruvate shifts the alternative pathway kinetics to the left allowing a greater electron flux at lower ubiquinone pool reduction levels. A reduction of disulphide bond, probably via matrix NADPH in a thioredoxin mediated reaction occurs as a result of enzyme activation. Pyruvate and some other 2 oxo acids (such as glyoxylate) stimulates the reduced enzyme which directly interacts directly with oxidase. The functions of AOX include optimization of respiratory metabolism, protection against excess of reactive oxygen species, adaptation to variable nutrition source and to biotic and abiotic stress factors.

A bacterial strain can use carbohydrates and hydrocarbons as growth substrates. The strain uses glucose following a minimal lag period aster inoculation, regardless of other carbohydrates and hydrocarbons in the growth medium. The following observations were also made:

- In the absence of glucose, lactose is used after a lag period of about three times as long as the lag period of glucose utilization
- The presence of hydrocarbons does not affect the lag period for the utilization of lactose.
- The utilization pattern for all hydrocarbons is similar to that of lactose
- Branched hydrocarbons are not immediately utilized if straight chain hydrocarbons are initially present.

A specific regulatory mechanism named diauxine is consistent with the above observation related to carbohydrate and hydrocarbon utilization.

Diauxic growth

If bacteria like *E.coli* are grown in a culture medium containing two carbon sources they exhibit a growth curve, called diauxic growth curve. In a diauxic growth, the *E.coli* first utilizes the glucose in the medium and after after completely exhausting glucose in the medium it starts to utilize lactose in the same medium. A short lag period is exhibited before feeding on lactose. From this we could conclude that *E.coli* preferentially utilizes certain carbon source.

Water and electrolytes like Na^+ and Cl^- are lost from the body in diarrhoea. Oral administration of NaCl solution in this condition does not improve the situation. When glucose is administered with normal NaCl solution through oral route, the absorption of electrolytes along with water occurs and the patient recovers because Na^+ is co-transported with glucose on the apical surface of the mucous cells of the small intestine.

Reason

Oral rehydration therapy (ORT) includes both sugar and salt. The consumption of salt and sugar helps to slow the evacuation of fluids allowing the absorption of electrolytes in the intestine. SGLT-1 transporter proteins are present in jejunum. Consuming Na^+ and glucose in the ratio of 2:1, SGLT-1 actively gets transported across epithelial wall thus creating an osmotic imbalance. This causes water to get immediately pulled into the vascular system which replenishes fluid and electrolytes instantly avoiding most of gastrointestinal tract. ORS and Zinc are more efficient.

The erythrocyte membrane cytoskeleton consists of a meshwork of proteins underlying the membrane. The principal component spectrin has α,β subunits which assemble to form tetramers. The cytoskeleton is anchored to the membrane through linkages with the transmembrane proteins band 3 and glycophorin C. The cytosolic domain of band 3 also serves as the binding site of glycolytic enzymes such as glyceraldehydes 3- phosphate dehydrogenase. Analysis of the blood sample of a patient with haemolytic anaemia shows spherical red blood cells. The patient carries mutant β spectrin defective in αβ dimerization ability.

Explanation

Spectrin has α,β subunits each of which are dimer. The dimers assemble to form tetramer thus giving shape to the principal component. It is given that the patient with haemolytic anaemia shows spherical red blood cells which means that the dimers are affected by the disorder and hence they cannot unite to form a tetramer.

A fixed smear of a bacterial culture is subjected to the following solutions in the order listed below and appeared red.

- *Carbolfuschin (heated)*
- *Acid alcohol*
- *Methylene blue*

Bacteria stained by this method were identified as Acid fast Mycobacterium species.

Reason

Acid fast stain (Ziehl Neelsen stain) is a special bacteriological stain used in the identification of *Mycobacteria*. This staining is much helpful in diagnosing *Mycobacterium tuberculosis* as it cell wall is rich in lipid thus making it unresponsive or resistant to gram staining. It is also used for staining some other bacteria like *Nocardia*. The reagents used in acid fast stain are carbolfuschin acid alcohol and methylene blue.

A researcher was studying a protein 'X' which has been observed to move across cells when an extracellular electrical stimulus is provided. An artificial peptide 'P' was prepared which resembles the structure of connexions and competitively inhibits connexion formation. The fate of protein 'X' if the cells are treated with peptide 'P' and then electrical stimulus is provided could be explained as: 'X' fails to move across due to improper formation of gap junctions

Reason

Connections also known as gap junction proteins are a family of structurally related transmembrane proteins which assemble to form vertebrate gap junctions.

Pyruvate dehydrogenase is subjected to feed back inhibition by its products in glycolysis. The chemical compounds involved in feedback inhibition of pyruvate dehydrogenase are NADH and Acetyl CoA.

Explanation

Feedback inhibition is a form of allosteric regulation. In feedback inhibition, the final product of a sequence of enzymatic reactions accumulates in abundance. As a result of formation of large amount of product, the final product binds to an allosteric site on the first enzyme in the series of reactions to inhibit its activity. NADH and Acetyl CoA are involved in glycolysis.

The auxotrophic strains of E.coli A (met bio⁻ thr⁺ leu⁺ thi⁻) and B (met⁺ bio⁺ thr⁻ leu⁻ thi⁻) were incubated together for 18 hours in a liquid complete medium and then - 10^8 cells were plated on a minimal medium. Phototrophs were observed at the frequency of 1×10^{-7} cells. This may have happened by a process of genetic recombination between the two strains or by mutation of strains. The control experiment that would help rule out the possibility of mutation is by growing strains A and B individually in a liquid complete medium for 18 hours and then plating them on a minimal medium.

Explanation

The mutant microbe that possesses a nutritional requirement not possessed by the parent is known as auxotroph. The possibility of recombination increases on growing strain A and B individually in liquid medium.

Addition of the antibiotic cephalexin to growing E.coli cells lead to filamentation of the cells, followed by lysis. Cephalexin is an inhibitor of peptidoglycan synthesis.

Explanation

Cefalexin is a beta-lactam antibiotic belonging to the cephalosporin family. It has bactericidal property and acts by inhibiting synthesis of the peptidoglycan layer of the bacterial cell wall. As cefalexin closely resembles d-alanyl-d-alanine, an aminoacid ending on the peptidoglycan layer of the cell wall, it can thus irreversibly bind to the active site of PBP, which is essential for the synthesis of cell wall.

Concepts related to Physiology

Key Note

Photosystem I and Photosystem II reaction centres are not uniformly distributed in the thylakoid membrane.

Reasoning of questions related to Physiology asked in CSIR UGC NET

Cells undergo apoptosis by two distinct and inter connected pathways: extrinsic and intrinsic. Intrinsic pathway is activated by extracellular ligand binding to cell surface death receptors. Whenever an apoptotic stimulus activates intrinsic pathway, the Pro apoptotic Bax and Bak proteins become activated and induce the release of cytochrome C from mitochondria leading to caspase cascade activation resulting in apoptosis. In cell A, cytochrome C is introduced by microinjection but Bax and Bak are inactivated. In both is cases, the apoptosis is induced and is high.

Explanation

Apoptosis otherwise known as programmed cell death is classified into two categories namely extrinsic and intrinsic. The binding of ligand to the cell surface activates the extrinsic apoptosis. In intrinsic apoptosis, the pro apoptotic Bax and Bak proteins are activated which induces the release of cytochrome C from mitochondria which leads to capsade cascade activation resulting in apoptosis. In the case of A, cytochrome C induced by microinjection and in the case of B, cytochrome C is introduced by microinjection and there is no need for the presence of Bax and Bok proteins in B. Thus in both A and B the apoptosis is induced and its rate is high.

Human chorionic gonadotropin (hCG) is known to facilitate attachment of blastocyst to uterus. In woman with mutation in hCG gene, biologically inactive hCG was formed but implantation occurred. When hCG was immune neutralized in the uterus of normal woman, implantation failed. This suggests that for implantation in humans, blastocyst can produce the required hCG, which help locally in uterine attachment.

Reason

hCG is secreted in syncytiotrophoblast by the placenta after implantation. This interacts with LHCG receptor of the ovary and promotes the maintenance of corpus luteum at the beginning of pregnancy. hCG functions by repelling the immune cell of mother thus protecting the fetus during first trimester. Hence when hCG is immunoneutralized it cannot perform the function of protecting the fetus.

An individual was suffering from digestive complications. It was observed that the individual had dehydrated gastrointestinal tract. When an advanced investigation was made the person was found to have defects in cystic fibrosis transmembrane conductance regulator protein.

Reason

Cystic fibrosis transmembrane conductance regulator (CFTR) is found in epithelial cells of many organs including lung, liver, pancreas, digestive tract and reproductive tract. CFTR is strongly expressed in the sebaceous and eccrine sweat glands in the skin.

The action potential was recorded intracellularly from a squid giant axon bathed in two types of fluid such as sea water and artificial sea water having lower concentration of sodium ions while maintaining the osmotic pressure with chlorine chloride. The nature of action potential was different in the two bathing fluids because the amplitude of the action potential was gradually decreased with the reduction of sodium concentration in bathing fluid but the duration of action potential was prolonged.

Explanation

Action potential is a nerve impulse generated by voltage gated Na channels that lasts for 1 millisecond and voltage gated calcium channels that last for 100 milliseconds or longer. Inward flow of sodium ions due to change in electrochemical gradient causes a further rise in membrane potential that opens many channels producing a great electric current across the cell membrane. There will be a rapid influx of sodium ions which cause polarity of the plasma membrane to reverse resulting in rapid inactivation of ion channels. When sodium channels are closed and the potassium channels are activated there will be an outward current of potassium ions, returning the electrochemical gradient to the resting state.

The following are statements about long distance translocation of photoassimilates in higher plants.

- *Sugars are translocated in the phloem by mass transfer along a hydrostatic pressure.*
- *Munch pressure flow hypothesis is crucial to drive translocation in the phloem.*
- *Allocation and partition of carbon within a source leaf determine the phloem loading phenomenon.*

Explanation

Translocation of sucrose to both roots and flowers from leaves is an example for bidirectional transport. Phloem loading is a process of loading carbon into the phloem for transporting it to different 'sinks' in plant. The transport starts from mesophyll cells to sieve elements. Phloem unloading is the transporting of sucrose from sieve elements to sink cells. Munch pressure flow hypothesis is most widely accepted hypotheis. Mass transfer of solute (sucrose in water) from source to sink occurs along a hydrostatic pressure gradient. As per Munch hypothesis, a high concentration of organic substance (sugar), inside phloem cells at a source like leaves create a diffusion gradient that draws water into the cell from adjacent xylem. This result in turgor pressure named hydrostatic pressure in the phloem.

An action potential was generated on a nerve fibre by a threshold electrical stimulus. When a second stimulus was applied, no matter how strong it was, during the absolute refractory period of the action potential, the nerve fibre was unable to generate second action potential. This observation could be explained as:

- A large fraction of sodium channels was voltage inactivated
- The critical number of potassium channels required to produce an action potential could not be recruited

Reason

At first, the inward flow of sodium ion rise the membrane potential leading to opening of many channels (electric current across membrane). As this closes potassium channels are activated leading to the outward current of potassium ions.

A patient has episodes of painful spontaneous muscle contraction, followed by periods of paralysis of the affected muscles. It was identified as primary hyperkalemic paralysis, an inherited disorder. The possible causes of the paralysis are elevation of extracellular K^+ which causes depolarization of skeletal muscle cells and sodium channels are voltage inactivated in depolarized state.

Explanation

Hyperkalemic periodic paralysis is an inherited autosomal dominant disorder which affects the sodium channels in muscle cells. It has the ability to regulate potassium levels in the blood. The patients exhibit increased level of potassium ions in their blood as a result of release of potassium ions into the bloodstream by the weak or paralyzed muscle. Mutations in sodium channels cause the inward flow of K^+ into the blood. This imbalance of K^+ ions results in difficulty in muscle contraction.

An EEG was recorded and its power spectrum analyses were done in rate with implanted electrode for a long time. The power of the EEG waves decreased two months after electrode implantation. This observation is due to

- Glial cells accumulate surrounding the exposed tips of electrodes

- Degeneration of neurons surrounding the electrode tips occur due to metal ion deposition

A nerve impulse or action potential is generated from transient changes in the permeability of the axon membrane to Na^+ and K^+ ions. The depolarization of the membrane beyond the threshold level leads to Na^+ flowing into the cell and a change in membrane potential to a positive value. The K^+ channel then opens allowing K+ to flow outwards ultimately restoring membrane potential to the resting value. The Na^+ and K^+ channels operate in opposite directions because there is a difference in Na^+ and K^+ concentrations on either side of the membrane.

Reason

Na^+ - K^+ pump is an electrogenic pump as more positive charges are pumped to the outside than to the inside (three Na^+ ions to outside for each two K^+ ions to the inside) leaving a net deficit of positive ions on the inside which results in a negative potential inside the cell membrane.

Statements related to seed development in plants

- During final phase of development, embryos of orthodox seeds become tolerant to dessication, dehydration, losing upto 90% of water.
- Precocious germination is the germination of seeds without passing through the normal quiescent and/or dormant stage of development.
- Absicic acid is known to inhibit precocious germination

Explanation

Orthodox seeds are the seeds that survive dry or freezing conditions during ex-situ conservation. These are long lived seeds and can be successfully dried to moisture content as low as 5% without injury and are able to tolerate freezing. These are also referred as dessication tolerant seeds. The longevity or the life span of these seeds is high. Recalcitrant seeds (Unorthodox seeds) are short lived seeds and cannot be dried to moisture content below 20-30% without injury and are unable to tolerate freezing. These are also called dessication sensitive seeds. The moisture content of the seeds resulting in microbial growth and are deteriorated. Quiescence is the state of suspended growth of embryo or the resting condition of seed. The seed is put into resting state (dessication). Quiescent embryos will resume growth at any time on exposing to favourable conditions like water, oxygen and warmth. Dormancy is a state that requires a special event or "trigger" before the embryo can resume growth such as fire, scarification or cold treatment. Precocious germination is the early germination of seed or embryo prior to full maturation of embryo. The seeds cannot be passed through dormant or quiescent stage. Precocious germination is inhibited by abscicic acid

Light is an important factor for plant growth and development. There are several photoreceptors in higher plants such as Arabidopsis thaliana involved in perception of various wavelengths of light. The statements related to photoreceptors:

- *Red light photoreceptors are represented by a gene family.*
- *Cryptochrome 1 and Cryptochrome 2 have evolved from bacterial DNA protolysases.*

Explanation

Photoreceptor proteins are light sensitive proteins. Phytochrome is class of photoreceptor that plants use to detect light. They are sensitive to light in the red and far red region of the visible spectrum. Cryptochromes are flavoproteins sensitive to blue light. Besides chlorophylls, cryptochromes are the only proteins known to form photoinduced radical pair invivo. Cryptochromes are derived from protolyases and are thus closely related to them. Protolyases are bacterial enzymes that are activated by light and are involved in the repair of UV induced DNA damage.

Carbohydrates synthesized by photosynthesis are converted into sucrose and transported via phloem to other parts of the plant. The following aspects are associated with sucrose uploading in phloem and its transport:

- *Sucrose uploading can be both symplastic and apoplastic*
- *The route of phloem uploading in mesophyll cells: phloem parenchyma-companion cells-sieve tubes.*
- *Transport in sieve tubes is as per the 'pressure flow model'.*

Explanation

Apoplasmic transport involves the transport of complexes containing high sugar concentration. By active transport, sugar is transported to phloem apoplasm. In symplasmic transport the turgor pressure of sieve elements and surrounding tissue decreases and the import of sugar rises. Example: Bulk flow of photoassimilates into root apex. Plants transform glucose into sugars before sending into phloem because the complex forms like sucrose and starch are more efficient in energy storage than the simple forms like glucose and fructose. Non reducing sugars do not have an OH group attached to the anomeric carbon and hence they cannot reduce other compounds. All monosaccharides including glucose are reducing sugars. A disaccharide can be either reducing or non reducing sugar. Maltose and lactose are reducing sugars, while sucrose is a non reducing sugar.

It is well established that "Band 3" protein of RBC membrane is solely responsible for Cl^- transport across membrane. A lysine group in the Cl^- binding site of 'Band 3' is crucial for this event. Keeping this is mind, the most appropriate way to load and retain a small anionic fluorescent probe (X) inside the RBCs suspended in phosphate buffered saline (PBs), pH 7.4 is to incubate the RBCs with X in Hepes sulphate buffer (pH 7.4) at 37°C for 30 minutes followed by treatment with NH_2 group modifying agent (Covalent modification).

Explanation

The idea here is to load and retain 'X' probe inside intat RBC. The lysine (positively charged) amino acid present inside the band 3 channel core, attracts Cl⁻ (negatively charged) and allows Cl⁻ accumulate inside. To retail all Cl⁻ inside just modify lysine NH_2 groups.

Concepts related to Immunology

Reasoning of questions related to Immunology asked in CSIR UGC NET

Mouse erythroleukemia (MEL) cells are used as an invitro cell culture model for understanding erythropoiesis. These cells are arrested at the stage of pro-erythroblast due to transformation. These cells are arrested at the stage of pro-erythroblast due to transformation. These cells could be induced by heme to differentiate further so as to synthesize haemoglobin. The most propable molecular mechansism for this could be that heme may suppress and / or down regulate an endogenous heme regulated inhibitor (HRI) Kinase, an inhibitor of globin synthesis. This down regulation inturn promotes differentiation

To validate this hypothesis the following approaches are appropriate

- ♠ Transfer of MEL cells with HRI kinase gene
- ♠ Knocking down of HRI kinase gene in MEL cells
- ♠ Measuring HRI kinase activity as a function of differentiation.

Dendritic cells (DC) from BALB/c mice were treated with IL-10 or with IFN-y. Similarly, dendritic cells from β_2- microglobulin deficient mice were also treated with IL-10 or with IFN-y. These cells were co cultured with CD8$^+$T cells from hen egg lysozyme (HEL)- specific T cell receptor transgenic mice in presence of the HEL peptide. Five days later, CD8+T cells were assayed for target cell lysis. Which one of the following combinations will have the highest target cytotoxicity

- ♠ DC(BALB/c) $^{IL\text{-}10}$ × CD8$^+$T
- ♠ DC (BALB/c) $^{IFN\text{-}y}$ × CD8$^+$T
- ♠ DC (β_2- microglobulin deficient) $^{IL\text{-}10}$ × CD8$^+$T
- ♠ DC (β_2- microglobulin deficient) $^{IFN\text{-}y}$ × CD8$^+$T

Answer

DC (β_2- microglobulin deficient) $^{IL\text{-}10}$ × CD8$^+$T

Reason

Dendritic cells (DC) are the antigen presenting cells. The function of dendritic cells is to process antigen and present it on the cell surface to the T- cells of the immune system. BALB/c is nothing but the albino mice used for research purposes. IL-10 is the human cytokine synthesis inhibitory factor (CSIF), is an anti inflammatory cytokine. Interferon gamma (IFN-γ) which is also known as immune interferon is a dimerized soluble cytokine. β_2- microglobulin is found on the surface of all nucleated cells. If it is absent very limited amount of MHC class I molecules can be detected on the surface. $CD8^+T$ is the subset of T cells involved in the development of acquired immunity. Cytotoxicity is the quality of being toxic to cells. Both DC(BALB/c) $^{IL-10}$ × $CD8^+T$ and DC (BALB/c) $^{IFN-\gamma}$ × $CD8^+T$ has β_2- microglobulin + $CD8^+T$. DC (β_2- microglobulin deficient) $^{IL-10}$ × $CD8^+T$ is $\beta2M$ deficient and an anti-inflammatory cytokine IL-10 is added to that. DC (β_2- microglobulin deficient) $^{IFN-\gamma}$ × $CD8^+T$ is also $\beta2M$ deficient but it is treated with the immune interferon. Hence when compared to all DC (β_2- microglobulin deficient) $^{IL-10}$ × $CD8^+T$ has highest target for cytotoxicity.

In bone marrow, stem cells are committed to different lineages. Factors that stimulate the colonies of these different lineages are interleukin 3 (multi- CSF), granulocyte- macrophage colony stimulating factor (G-CSF or M-CSF). In a mouse deficient in GM-CSF, the number of hematopoietic cells will be altered because mast cell will be normal in number while granulocytes and macrophages will be deficient in number.

Reason

Granulocyte- macrophage colony- stimulating factor (GM-CSF) is a 23 kDa glycoprotein. It is produced by many cell types including monocytes or macrophages, B lymphocytes, neutrophils, eosinophils, pulmonary type II cells and other respiratory epithelial cells. GM-CSF performs a variety of functions like regulation of hematopoietic cell proliferation and differentiation and modulation of the function of mature hematpoeitic cells. Their effects include enhanced antigen presentation, increased complement and antibody mediated phagocytosis, augmentes microbicidal capacity and heightened leukocyte chemotaxis and adhesion.

A set of neonatal mice are divided into four groups. Group 1 neonates were not primed with any antigen. Group 2 neonates were primed with KLH. Group 3 neonates were primed with KLH but thymectomized. Group 4 neonates were KLH- primed, thymectomized, but reconstituted with KLH-specific CD4+T cells. All these mice when grown adult were challenged with KLH and anti KLH IGC antibody was measured in sera. The correct order of magnitude of antibody response

    Group 4> Group 1> Group 2 ≥ Group 3

Reason

It is given that group 1 is not primed with antigen. Group 2 is primed with KLH/ Keyhole limpet hemocyanin. KLH is an oxygen carrying metalloprotein, in the presence of oxygen it turns blue and in the case of absence of oxygen it becomes colourless. KLH is immunogenic, but doesn't cause an adverse immune response in human. It is used as vaccine carrier protein. Group 3 is primed with KLH but thymectomized i.e. it is operated and removed thymus (closely associated with immune system).Group 4 is KLH primed, thymectomized but reconstituted with KLH specific CD4+T cells which is a glycoprotein found on surface of immune cells such as T helper cells. Therefore group 4 has antigen and makes the appearance of more antibody. Group 1 is normal and body antibody will be present. Group 2 has KLH and Group 3 has KLH but lack thymus.

One highly pathogenic DNA virus enters into the host cells by endocytosis replicates in the nucleus followed by cell lysis. To prevent this infection the drugs that cause acidification of vesicle and nuclear export are used.

Explanation

The necessary step in many processes like receptor recycling, virus penetration and entry of diphtheria toxin into cells is the acidification of endocytic vesicles. The export of proteins, tRNAs and assembled ribosomes is termed as nuclear export. Hence to prevent infection by pathogenic DNA virus acidification of vesicles and nuclear export must be blocked.

A BALB/c mouse was thymectomized on the first day after birth (mouse 1) whereas another was thymectomized on day 7 after birth (mouse 2). A third mouse underwent the same operation on day 21 after birth. After 56 days, sera were prepared from these mice and also from control mice, which had sham operation. The sera were checked for anti DNA antibodies it was observed that only mouse 1 had anti DNA antibodies.

Reason

In a study, thymectomization i.e. the operation for removal of thymus was carried out in a new born mouse in less than 24 hours after birth. The goal of the study was to remove the source of abnormal antibody production causing the disease thus leading to the resolution of symptoms. The mouse was then immunized with intra peritoneal injection of sheep erythrocytes at 6 weeks of age. Then the serum haemolysin activity was measured. The reduction in circulating antibody activity and delay in reaching maximum levels are obtained.

A potentially valuable therapeutic approach for killing tumour cells without affecting the normal cells is the use of immunotoxins. Immunotoxins consists of particular cell specific monoclonal antibodies coupled to lethal toxins. The following molecular approaches are appropriate for killing the tumour cells.

- ♠ Cell surface receptor binding polypeptide chain of toxin molecules should be replaced by monoclonal antibodies which are specific for a particular tumour cell

- ♠ *Constant region Fc domain of tumour cell specific monoclonal antibody should be replaced by toxin molecules.*
- ♠ *Inhibitor polypeptide chain of toxin molecules should be conjugated to the F(ab) domain of tumour specific monoclonal antibody.*

Reason

A human made protein with targeting portion linked to toxin is termed as immunotoxin. When this protein binds to cell it is taken in by endocytosis resulting in death of the cell. These chimeric proteins are usually made of a modified antibody or antibody fragment, attached to a fragment of a toxin. The targeting portion is composed of the Fab portion of an antibody. The toxin used is a cytotoxic protein derived from a bacterial or plant protein, from which the natural binding domain has been removed so that the FV directs the toxin to the antigen on the target cell.

Influenza virus (IV), a well known enveloped animal virus, enters its host cells through membrane fusion process catalyzed by haemagluttinin (HA) protein inside endosomes at 37°C. HA is localized in the lipid bilayer membrane of the IV as an integral membrane protein and is responsible for binding and fusion of IV membrane with the endosomal membrane of the host cells. Upon binding, IV is internalized into host cells through receptor mediated endocytosis followed by fusion of the IV membrane with endosome membrane catalyzed by HA. In a situation, if we wish to fuse IV membrane with its host cells (deficient in endocytosis) at the plasma membrane, IV and host cells are allowed to bind and fuse at pH 5.0 and 37°C.

Reason

Haemaaglutination takes place in acidic pH.

Glycophorin of RBC membrane spans the membrane only once and the N-terminal is projected extracellularly and the C-terminal is exposed to the cytosolic side. With the help of antibodies (labelled with fluorophores) against N-terminal and C-terminal peptides, orientation of glycophorin across membrane can be verified. The correct statement derived from the above is permeabilizeation is required for C-terminal antibodies as well as N-terminal antibodies

Reason

Permeabilization is required for C-terminal antibodies to get inside. The inside out host contains N-terminal inside and C terminal outside.

Concepts related to Evolution

Some Terms and concepts Explained

ADAPTIVE RADIATION

Adaptive radiation is a process by which the organisms diversify rapidly from an ancestral species into multitude of new forms.

DIVERGENT EVOLUTION

In divergent evolution accumulation of differences between groups can lead to the formation of new species is usually due to diffusion of the same species to different and isolated environments. Thus the block the gene flow among the distinct population is prevented.

CONVERGENT EVOLUTION

In convergent evolution, the organisms which are not closely related evolve similar traits independently as a result of adapting to similar environment and ecological niche.

MICROEVOLUTION AND MACROEVOLUTION

The change in allele frequencies that occur over time within a population is termed as microevolution. Microevolution occurs as a result of one of the five processes namely mutation, selection (artificial and natural), gene flow, gene migration and genetic drift. This change occurs over a relatively short amount of time. Microevolution over time leads to speciation or the appearance of novel structures sometimes grouped as macroevolution. Microevolution refers to a small evolutionary change within a species or population whereas macroevolution is the evolution on a large scale including that of separated gene pools. Macroevolutionary studies focus on changes occurring at or above the level of species.

Reasoning of questions related to Evolution asked in CSIR UGC NET

The Galapagos finches were an important clue to Darwin's thinking about the origin of species. These finches are believed to have descended from a single ancestral species that colonized the Galapagos Archipelago, America, over a short period of time. The Galapagos finches differ in their beak shape and size. Different species feed on seeds that vary in size and hardness because the finches represent an example of adaptive radiation in which beak variation was generated by mutation followed by selection of different seed types.

Explanation

Adaptive radiation is a process in which organism diversify rapidly from an ancestral species to form a multitude of new forms. This occurs when a change in the environment makes new resources available creating a new challenges or open new environmental niche.

In order to demonstrate that the long tails of males attracted females in a bird species, experiments captured and cut the tails of 'n' number of males and monitored the number of females mated by each male. They had two types of controls in the experiment.

   1) 'n' males were not captured
   2) 'n' males that were captured, had their tails cut and then stitched back to attain the original size
   The males with cut tails mated with a significantly smaller number of female than both the controls.

Reason

The stress of cutting tails affected the performance of males. The time wasted in the capture reduced the mating opportunities of males. Female chose males randomly. But it is well understood that females did not avoid any deviation from normal since they had mated with cut tailed males.

Fore limb of human and flippers of whale are embryologically homologous structures. The study of homologous structures tells us the following about evolution

- ♠ This is an example of adaptive radiation occurred due to similar group of organisms inhabiting different environments.
- ♠ This is an example of divergent evolution, occurring due to similar group of organisms inhabiting different environments.
- ♠ Similar group of organisms with mutations and variations getting naturally selected in different environments.

Wolbachia are obligate intracellular bacteria, many different strains of which are abundantly present in insects. They induce mating incompatibility in host, (i.e) males infected with one strain can one strain can only fertilise females infected with the same strain. No other pathological effects are observed in host.

A possible evolutionary consequence of this phenomenon would be reproductive isolation leading to a rapid speciation in an insect.

Twenty small populations of a species, each polymorphic for a given locus (T,t) were bred in captivity. In 10 of them the population size was kept constant by random removal of individuals, while other 10 were allowed to increase their population size. After several generations it was observed that in 7 of the size restricted populations only T was present, in the remaining 3 only t was present. In the growing populations 8 retained their polymorphism and in 2 only (t) was observed.

The experiment illustrates that the genetic drift is more likely observed in small population.

Concepts related to Embryology

Some concepts Explained

DEVELOPMENT OF SEA URCHIN EMBRYO

Sea Urchin exhibit radial holoblastic cleavage. The 1st and 2nd cleavage is meridonial and perpendicular to each other. This means that the cleavage furrows pass through the animal vegetal pole. The 3rd cleavage is equatorial and is perpendicular to the 1st and 2nd cleavage. This division separates the animal poles from one another. As a result of 4th cleavage, 4 cells of the animal tier divide meridonially into 8 blastomeres of same volume which are called mesomeres. Vegetal tier exhibits unequal equatorial cleavage with four large cells- macromeres and four small cells – micromeres. The 16 celled embryo cleaves into 8 mesomers which divide to produce 2 animal tiers (an 1& an 2). The 6th division results in the formation of 128 celled blastula. Although the early blastomeres have specific fates in the larva, majority of the fate is achieved by conditional specification. The only cells whose fate is determined autonomously are the skeletogenic micromeres.

STEPS IN VULVA DEVELOPMENT

VPCs Generation

During L_1 and L_2 stages, 6 vulval precursor cells (VPCs) are generated from the cells that are located in the ventral epidermis.

Patterning of Vulval precursor

During the L_3 stage, a signal from the gonad and the signals among the VPCs specify 3 VPC resulting in generation of vulval cells. The vulval lineages are of 2 types: 1° and 2° each of which generates a distinct set of progeny. The non induced VPCs generate a 3° lineage, which makes up the epidermal cells that fuse with the large syncytial epidermis hyp7.

Adult cells Generation

The adult vulva is comprised of 22 nuclei in cells which are of 7 types.

Anchor cell invasion

The anchor cell extension progresses to the centre of the VIIF cells to form a hole in the epidermis.

Vulva morphogenesis

The seven types of vulval cells invaginate and sequentially form seven distinct toroids, connecting to the uterus and everts as the hermaphrodite changes to adulthood. Anchor cell is specified from among two somatic gonadal cells during the end of L_2 stage. The anchor cell also patterns the developing uterus, including the six Pi cells responsible for generating the utse cell and UV1 cells. The UV 1 cells attach to the VulF cells. The anchored cell ultimately fuses with the utse.

Reasoning of questions related to Embryology asked in CSIR UGC NET

Polyspermy results when two or more sperms fertilize an egg. It is usually lethal since it results in blastomeres with different numbers and types of chromosomes. Many species therefore have two blocks to polyspermy: the fast block and slow block. In the case of sea urchin, fast block is immediate and causes the egg membrane resting potential to rise which does not allow the sperm to fuse with the egg and is mediated by an influx of sodium ions. Slow block or cortical granule reaction is mediated by calcium ions.

Explanation

Prevention of polyspermy depends on The change in electrical charge across the egg as a result of fusion of sperm and egg prevents polyspermy. In Fast block polyspermy the ionic concentration of egg differs slightly from the surrounding (sodium and potassium ions). Sea water has high sodium ion concentration whereas egg cytoplasm has low level of sodium ion concentration. Reverse in the case with potassium ion concentration that is, its concentration is low in sea water and high in egg's cytoplasm. The fast block polyspermy is achieved by lowering sodium ion concentration in the surrounding sea water. The slow block polyspermy is carried out by removing the vitelline membrane surrounding the sperm.

In an experiment, the cells that would normally become the middle segment of Drosophila leg was removed from the leg forming area of the larva and were placed in the tip of the fly's antenna. Based on the "French flag" analogy for the operation of a gradient of positional information, the transplanted cell retains their committed status as leg cells but respond to the positional information of their environment by becoming leg tip cells (i.e. claws).

Explanation

French flag analogy was proposed by Lewis Wolpert. This model uses French tricolor flag as visual representation to visualize the interpretation of genetic code by embryonic cells to create the same pattern even after removing certain pieces of embryo are removed. When a portion of a positional field is removed, the remaining cell's positional values change, as with their genetic identities thus permitting regeneration.

A set of experiments that were carried out to demonstrate the effect of apical ectodermal ridge (AER) of the chick limb bud on the underlying mesenchyme are enlisted below along with their expected outcome:

- *Removal of AER of forelimb leads to cessation of limb development*
- *If an extra AER is placed in the forelimb bud, duplication of the distal region of the wing takes place.*
- *If AER of forelimb bud is replaced with beads soaked in FGF 2, a normal wing develops.*

Explanation

The development of limb halts on removing AER. When an FGF bead is added in the place of AER, normal limb development proceeds. On adding an extra AER, two limbs formation occurs (duplication). When forelimb mesenchyme is replaced with hindlimb mesenchyme, a hindlimb grows. When forelimb mesenchyme is replaced with non limb mesenchyme, the limb development halts on the regression of AER. When AER from a late limb bud is transplanted to an earlier limb, the limb formation occurs normally.

Hensen's node is established as the avian equivalent of the amphibian dorsal blastophore lip. The following observations are presumed to be in support of the same

- *It is the region whose cells can both induce and pattern a second embryonic axis when transplanted into other locations of the gastrula.*
- *It expresses the same marker genes as the Spermann's organizer in amphibians.*

Reason

The site of beginning of gastrulation is the node. Gastrulation is the process of formation of three germ layers. The determination and patterning of the anterior- posterior axis of the embryo (embryonic development) is concerned with the node.

Development of vulva in C.elegans is initiated by the induction of a small number of cells by short range signals from a single inducing cell. With reference to this, the following statements were put forward

- *When the anchor cell was ablated early in the development, no vulva formed.*

♠ *A constitutive signal from the hypodermis inhibits the development of both primary and secondary fates but it is overruled by the initial signal from the anchor cell.*

Explanation

C.elegans is a hermaphrodite in which the vulva develops from the ventral epidermal precursors during post embryonic (larval) development. In this case the vulva connects the developing uterus to the external environment. In the case of adult, the vulva is needed for egg laying and copulation with males.

Instructive and permissive interactions are the two major modes of inductive interactions during development. The following compares some properties of cell lines and cord blood stem cells. Cell lines, which are stored in liquid nitrogen, can be retrived for experiments where they behave as per their original self. Cord blood can also be retrived from liquid nitrogen for procuring stem cells. Unlike cell lines, the stem cells can be additionally induced to undergo differentiation into desired lineages, which are very different from the original self. The behaviour of cell lines and stem cells is analogous to which of the interactions?

Answer

Cell lines exhibit permissive interaction whereas stem cells exhibit instructive interaction.

If you remove a set of cells from an early embryo, you observe the adult organisms lack the structure that would have been produced from those cells. Therefore, the organisms seem to have undergone autonomous specification.

Reason

Early embryo cells do not possess any varied cells and exhibit autonomous specification.

What would happen as a result of a transplantation experiment in a chick embryo where the leg mesenchyme is placed directly beneath the wing apical ectodermal ridge (AER).

Answer

Distal hindlimb structures develop at the end of the limb.

Driesh performed the famous "pressure plate experiment" involving intricate recombination with an 8-celled sea urchin embryo. This procedure reshuffled the nuclei that normally would have been in the region destined to form endoderm into the presumptive ectoderm region. If segregation of nuclear determinants had occurred, the resulting embryo should have been disordered. However, Driesch obtained normal larvae from these embryos. The possible interpretations regarding the 8-celled sea urchin embryo are:

- *The prospective potency of an isolated blastomere is greater than its actual prospective fate.*
- *Sea-Urchin embryo is a 'harmoniously equipotent system' because all of its potentially independent parts interacted together to form a single embryo*
- *Regulative development occurs where location of the cell in the embryo determine its fate.*

Reason

Driesch had demonstrated that the prospective potency of an isolated blastomere is greater than its prospective fate. According to Weismann and Roux, the prospective potency and the prospective fate of the blastomere must be identical. Driesch concluded that the embryo of the sea urchin is a 'harmoniously equipotent system' because all of its potentially independent parts function together to form a single organism. He also concluded that the fate of a nucleus depends on its location in the embryo

Concepts related to Systematics and Ecology

Some Terms and concepts Explained

COELOMATE

Coelom is the main body cavity and is positioned inside the body to surround and contain the digestive tract and other organs. In developed animals, it is linked with a mesodermal epithelium. In Molluscs it remains undifferentiated. Coelomates are grouped into three categories: True coelomate, Pseudo coelomate and Acoelomate. True coelomates have a body cavity called a coelom with a complete lining called peritoneum derived from mesoderm. In Pseudo coelomates, the tissue from mesoderm only partially lines the body cavity. Acoelomates (flatworms) have no body cavity. Semisolid mesodermal tissue between the gut and body wall hold their organs in place.

SYSTEMATIC HIERARCHY

- Kingdom
- Phylum
- Subphylum
- Super class
- Class
- Sub class
- Cohort
- Order
- Sub order
- Super family
- Family
- Sub family
- Genus
- Sub genus
- Species
- Sub species

Models of succession

Facilitation model

Species with qualities ideal for 'early succession' colonizes the newly exposed landform after an ecological disturbance. The qualities include highly effective method of dispersal, the ability to remain dormant for a long period of time and the rapid growth rate. If the area has been heavily populated by the surrounding species, the pioneer species are often less due to shade and litter. Thus the presence of early succession species often changes the environment so that the habitat is less hospitable for the original species and facilitate invasion of later succession species.

Tolerance model

According to this model, the new pioneer species neither inhibit nor facilitate the growth and success of other species. The dominant species replace or reduce pioneer species abundance through competition eventually.

Inhibition model

According to inhibition model the earlier succession species actually inhibits the growth of later succession species and also reduces the growth of colonizing species that are already present.

Lindeman's efficiency

Lindeman's efficiency between trophic levels is depicted by the formula

Efficiency = A/B

Where A and B respectively are assimilation at trophic level n and assimilation at trophic level n-1 respectively

Reasoning of questions related to Systematics and Ecology asked in CSIR UGC NET

As per ICBN 2006 Agrostis stolonifera L. × Polypogon monspeliensis (L) Desf. is a Nothospecies.

Explanation

A nothospecies is a hybrid which is formed by direct hybridization of two species. It is indicated by a multiplication symbol (×) in between parents' binomial species names (when both parent species are known).

A species has the following population characteristics

- *Reduction in population size greater than or equal to 90% over the last 10 years or 3 generations*
- *Geographic range: Extent of occurrence: less than 100 Km2 and area of occupancy: less than 10 Km2*
- *Population size less than 50 matured individuals*
- *Probability of extinction in the wild is atleast 50% within the next 10 years or 3 generations*

Based on the above details we could formulate that the species belong to the critically endangered category.

Explanation

The critically endangered species are the species facing a very high risk of extinction in the wild and are in the verge to extinct any moment in the immediate future. The extinction is due to climate change, habitat destruction etc.

Reference

Questions referred from the official website of Human Resource Development Group (csirhrdg.res.in)

www.ingramcontent.com/pod-product-compliance
Lightning Source LLC
Chambersburg PA
CBHW070217230526
45471CB00002B/968